FORAGING

The Ultimate Beginners Guide to Foraging Wild Edible Plants and Medicinal Herbs

Jonathan S. Hunt

TABLE OF CONTENTS

INTRODUCTION

NATURE'S GIFT

The ability to forage for wild herbal plants to utilize their medicinal and nutritive properties is a gift. Even if you only master the humble Dandelion, you will be rich with the knowledge of a wild plant packed with vitamins, minerals and healing properties. This book can assist anyone who is on their journey to introducing the healthful benefits of wild herbs into their daily lives.

This guide has been created to bring you the power of health-giving wild medicinal herbs. You will learn everything you need to know about foraging common medicinal herbs, how to make the most of their healing properties and how to preserve them for later use. Also included are simple instructions on how to identify 10 of the most recognizable wild herbs available along with easy instructions on how to tap into their healing properties and how to make the most of their nutritional value. The purpose of this book is to introduce a new way to increase your vitality by using the earth's natural bounty.

How to Use This Guide

Though the wild medicinal herbs listed in this book are easily identifiable, it is always essential to have a local wildcrafter or botanist show you in person which herbs are which. Medicinal herbs should never be picked unless you are 100% sure of the species. Usually once you have harvested a wild plant in person, you won't forget.

The herbs described below have many beneficial effects but checking with your healthcare practitioner before taking any herbs or wild plants is extremely important. There are some herbs, such as Saint John's Wort, that can be dangerous when mixed with certain pharmaceuticals. And there are poisonous plants out there. White Berry Bane, which has cardiogenic properties, can actually cause cardiac arrest if many of its berries are eaten.

Of course, when eating wild plants our sense of taste, smell and even sight can often warn us of danger. Plants that are too high in tannins for human consumption (such as wild, uncooked acorns) will cause so much discomfort when first tasted that it's almost impossible not to spit them out. Many toxic plants actually look toxic, like the Berry Bane berries mentioned above which resemble alien eyeballs. Bright red and white colors will often be a signal to you that a plant is poisonous, but not always. Because of the inherent danger of ingesting some herbs, it is smart to be careful and safe. The best way to be sure about the exact identification of an herb is to have a local professional show you.

To get the most out of the information in this book, start by foraging the herbs you already know by sight; most people already know what a red clover looks like, for instance. Read the corresponding chapter in this book to learn about its medicinal properties, uses and how to harvest it. Remember to check with your healthcare practitioner before trying out the methods utilizing your foraged herbs. Find other people in your area who are interested in foraging; herbalists, naturalists, botanists and forestry experts are often your best bets for accurate wild herb identification.

Are Wild Plants Better?

All of the herbs discussed in this book can be bought in bulk in just about any natural foods store or online. There are many reasons why foraging wild plants is better than buying processed herbs. Knowing the source of your herbs empowers you to put only the purest of medicines into your body. Picking wild herbs is certainly more economically viable than buying them. But the most important reason to choose a wild plant over a store-bought herb is freshness. An herb picked straight from the ground still has its full power. It hasn't been dried, heated, waxed or sprayed nor has it been exposed to exhaust, refrigeration, irradiation, pesticides or other people's germs. Minimal processing of wild herbs can help retain their freshness and when done personally, the peace of mind is there; you know exactly where that plant lived and what conditions it's been subjected to since being picked.

The Benefits of Foraging Wild Medicinal Herbs

There are so many reasons why foraging for herbs is beneficial. One is that the act of foraging in itself is therapeutic. It creates an opportunity to get out in the wild, connect with the earth, get some exercise and enjoy fresh air and sunshine. Another important benefit of using wild plants is that the herbs you are foraging not only have medicinal properties, but many are also nutritional powerhouses. As an example, one cup of chopped dandelion greens provides 112% of the daily-recommended amount of Vitamin A, 32% of Vitamin C, 10% of Calcium and 9% of Iron. Additionally, the medicinal properties of these plants can be an excellent alternative to pharmaceuticals; oftentimes the only side effects noticed are positive, such as improved healing and increased vigor.

Understanding Ethical Wildcrafting

It's important to tread lightly when foraging for medicinal plants. Some herbs that have become mainstream have gotten over-harvested and their species are threatened. So it is important to only forage when there is plenty and take just what is needed for your personal use. There are methods of harvesting that mimic pruning and actually enhance the growth of a plant. As a general rule of thumb, when cutting the flowering tops of an herb, cut off about 5 or 6 inches from the top, leaving plenty of growth opportunity in what remains of the plant. When harvesting flowers, never take them all; leave plenty of seeds to ensure the herb survives. When digging up roots, slicing off a portion of the root and returning it to the ground is one way of replanting that herb. The wild medicinals are there for the taking, but it's essential that they remain there for our children and our children's children.

The Best Time to Forage

Generally, when harvesting flowers and leaves, the best time of the year to collect them is when they are fresh and new, which varies from plant to plant depending on their growth cycle and the part of the plant you are foraging. Leaves are best picked before the plant has flowered but can be picked after, they just may have lesser potency. Flowers can be picked when they are vibrant and fairly new. Roots may be harvested in early spring or fall. By foraging the roots before the greens and flowers have sprouted, or after they've browned and died away, you will be able to extract the strongest medicinal actions out of your harvested roots.

Ideally, leaves and flowers should be cut or picked around 10:00 a.m. so that the morning dew has dried off but they are not yet sapped of their power by the hot sun. On a cloudy day herbs can be harvested later. For many plants, once flowering has occurred the taste of the leaves becomes bitterer and their power is decreased. For roots, the time of day is less of an issue since they are underground and usually damp, thus not affected by the sun. These are basic guidelines. More information for each particular herb will be offered in the medicinal plant chapters.

Using Common Sense When Foraging

Good places to look for wild medicinals include meadows and forests that are not sprayed or littered. If you have a back yard with grass, fields or woods, chances are you can find medicinal herbs right on your own property. Knowing the health of the land from which you are foraging is important. Plants that are close to roads are not optimal, nor those near runoffs. As always, it's extremely important to get permission before foraging on someone else's property.

CHAPTER 1

TEN UNMISTAKABLE WILD MEDICINAL HERBS

We begin this journey into foraging medicinal herbs by studying some of the most widely recognized and easily identified wild plants. Along with information on how to find and harvest these herbs, you will learn their medicinal properties, their nutritional value and how to use these herbs to improve your health.

Dandelion – Taraxacum

All parts of the humble and versatile dandelion have been used extensively in herbal healing. The young leaves (present before the plant flowers) have diuretic properties and are rich in potassium, making them a favorable alternative to pharmaceutical diuretics which often leach potassium from the body as a side effect. The greens of the Dandelion make an excellent bitter tonic, stimulating bile flow and acting as a general kidney and urinary tract tonic.

The bitterness will likely increase tenfold after the plant flowers. While the tender young greens are enjoyable eaten raw, the older greens may be mixed with other greens in a salad, or dried and preserved for adding into teas, tinctures, soups or smoothies. Steaming any of the leaves with a pinch of lemon juice makes for a nutrient packed treat. Both the young and the old leaves can be made into a tea, either fresh or dried.

Dandelion roots can be ground and made into tinctures (see chapter 3 for how to make teas and tinctures) or even used as a coffee substitute. The powdered root makes an excellent tonic for digestive issues and skin conditions.

The flowers of the wild Dandelion are edible fresh or cooked and, in addition to similar properties to that of the roots and leaves, they act as an antioxidant. Traditionally it is common to make a wine with Dandelion flowers to tone the digestive system, kidneys and liver, making the perfect spring tonic. Finding Dandelion in the wild is quite easy - look on someone's untreated lawn. It loves wide open areas and often favors moist ground.

Wild Dandelion is often one of the first herbs to show itself in the spring and offers a perfect way to fortify our winter bodies.

Mullein-Verbascum Thapsus

Quite possibly the most obvious wild plant to identify, Mullein offers many uses for those foragers seeking remedies for respiratory ailments and ear infections. The large, thick and fuzzy leaves, though edible, are not the best tasting wild green and have a strange texture. But when made into a tea, Mullein leaves work hard as an expectorant and mucus membrane tonifier. They also have mild diuretic and sedative properties. Perfect for relieving coughs and colds, Mullein leaves can be used fresh or dried and make a soothing, cup of tea packed with Vitamins B2, B5 and B12.

Mullein flowers, when infused in oil, are used to relieve ear infections or other ear problems. A few drops of the Mullein oil in each ear and plugged gently with a cotton ball several times a day will bring relief to those suffering ear pain. The flowers are very small and since you need a lot to make an infusion, be sure you are picking where there are multiple Mullein stalks and make sure some flowers are left behind to self sow.

As a biennial plant, Mullein appears during its first year as a furry, whitish-tinted floret of leaves that hugs the ground. In the second year a long stalk will grow up through the middle of the floret, sometimes reaching upwards of 6 feet tall or more with leaves growing up along the stalk on the way. At the top of the stalk the tiny yellow flowers will appear in mid to late summer. In the fall the stalks can be harvested, coated with oil and used as outdoor lanterns. Mullein can be found in many disturbed bits of land, often growing up out of recently dug gravel and is prominent throughout the northern hemisphere.

Broad Leaf Plantain-Plantago Major

Another so-called weed that many people seek to eradicate from their lawns, wild Plantain presents the perfect outdoor remedy for cuts, stings and insect bites. This prolific herb offers up its medicinal value through cracks in the sidewalk, under a playground slide or almost anywhere else, especially enjoying a lovely green lawn. The leaves, when crushed, can be directly applied to external wounds and bites. Plantain contains astringent, antibacterial and anti-inflammatory properties and contains allantoin, a chemical that helps with new cell growth and tissue regeneration.

Internally, Plantain's demulcent properties help soothe digestive tract disorders and inflamed membranes, often aiding in coughs as well. And as if that weren't enough, the thin seed stalks, or husks, contain none other than Psyllium seeds. As an excellent and nutritional source of fiber, and heavy with mucilage, Psyllium seeds offer gentle healing of digestive disorders, diarrhea, and constipation.

When harvesting Plantain, cut the youngest leaves, which grow during most seasons, and use them immediately or let dry for a week out of the direct sun and in a spot with good air circulation. Seed stalks can be harvested from about mid-summer onward, when they are tall and green. Once dried, Plantain can be stored and used later for teas, healing washes for wounds, tinctures and salves.

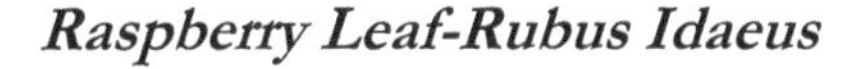

Raspberry Leaf-Rubus Idaeus

Loaded with vitamins and minerals, including potassium and calcium, Raspberry leaves makes a premier tea (however, it is highly advised not to use it during pregnancy). In addition, with their anti-inflammatory powers, a tea of the leaves taken 3

times daily can have a profound effect on arthritis and joint pain. Their astringent and antiseptic properties promote immunity internally and wound healing when used externally. As for the berries, one cup of the berries will provide over half of your daily Vitamin C and a third of the recommended daily dietary fiber.

Hard to miss in the wild, and easily recognizable, it's best to seek out the berries first. Once you know a patch in the wild, enjoy the nutrient packed berries. Then next spring you can take cuttings of the leaves, since the younger the leaves are the better. You can certainly harvest leaves once the berries are present, but as with most wild greens, it is ideal to harvest the young leaves.

Raspberries often can be found in areas that have been forested within the past 2-5 years. If wildcrafting in such an area, be aware that these places of new growth are often popular with bees which may have hives in the loose piles of undergrowth. Usually you will hear their sounds before getting too close, but it's good to keep that in mind. Wearing long sleeves and sturdy pants (like jeans) will help prevent you from getting scratched by the prickles.

Raspberry leaves are high in tannins, which makes them one of the easier and faster herbs to dry. Because the leaves are prickly, it's a better idea to dry them before use. Once they are dried they are more crumbly and the thorns seem to lose their bite. Large amounts of the tea can be drunk regularly with no ill effects and the cooled tea also doubles as a wash for wounds, bites and stings.

Red Clover-Trifolium Pratense

Commonly found throughout the northern hemisphere, Red Clover is another of those little weeds that people are always trying to rid from their lawns. If they only knew the extensive list of medicinal uses of this little flower! To start, Red Clover flowers make the gentlest of healing teas, appropriate for all ages and long term use for soothing nerves and easing bronchial ailments. The list of medicinal properties here is long: they act as an alternative for blood cleansing, an antispasmodic and expectorant for coughs and lung problems, a diuretic, and much more. Using Red Clover tea or tincture over a period of time will act as an overall cleanser and tonifier for your body, often relieving tough cases of eczema and psoriasis. Externally the cooled tea, once cooled, can be applied to wounds.

Red Clover is also a great source for isoflavones, which gently mimic estrogens making this plant excellent for menopausal symptoms. Packed with minerals like potassium and magnesium and many nutrients including Vitamin C and Thiamin (B1), a tea with the flowers of Red Clover offers optimal nutrition along with its healing effects.

When harvesting Red Clover, simply pluck the flowers off when they are at their brightest in color, having just bloomed. It's good to let them dry outside for a few hours in the shade before bringing them into the home. Like many clustery flowers, Red Clover makes an excellent home to tiny insects that will crawl out once the plant had been disturbed. Leave plenty of flowers for the bees and other pollinating insects that love this little plant.

Saint John's Wort-Hypericum Perforatum

There are many small yellow flowers in the wild. Only Saint John's Wort has tiny dark red spots underneath their small pointy petals that when touched, will temporarily stain your fingers a deep red. These same little dots will cause an oil infused with St. John's Wort to turn a gorgeous, burnt red color. The oil itself makes the perfect remedy for burns and a healing massage oil for sore muscles, pain and inflammation. Later in the book there are specific instructions on making infused oils.

St. John's Wort grows throughout the northern hemisphere particularly in meadows, at the edge of forests and in sandy soil conditions. The top 6 inches or so of the flowers, stems and leaves can be cut, dried, ground, capsulated and taken daily for depression, nervous distress, and nerve pain. Drinking a tea of the dried herb three times a day would also cause a positive effect on any nervous disorders.

Traditionally picked on St. John's Day, June 24th, throughout England and Scotland, this herb will bloom a bit later in northern U.S. climates; about two weeks after the 24th. Often considered another pesky plant, St. John's Wort's growth can become prolific.

There have been studies concluding that St. John's Wort causes sun sensitivity, but only on grazing animals eating copious amounts of the herb. Studies on humans have not revealed this problem.

Though Saint John's Wort is widely used for depression, it has been shown to cause serious health issues when mixed with certain pharmaceuticals and caution must be taken. Consult your healthcare practitioner before taking any of these herbs. Using St. John's Wort externally, as in applying infused oils directly onto burns or massaging the oil into the skin for pain, will not cause that interference with drugs.

Sheep Sorrel-Rumex Acetosella

This versatile little weed is easy to recognize, though it is somewhat small, rarely growing higher than 12 to 16 inches. The bright green spear-shaped leaves make a delicious treat to eat raw, add to salads or dry and preserve for teas. They have a tart taste, similar to that of a sweet tart, making it one of the more delicious wild greens. The tiny flowers that grow up on the thin stalks start out a creamy color and eventually turn a rich red. This is one example of wild greens that do not turn bitter after the plant flowers. A lover of grass, meadows and your garden, this herb is often mowed down or pulled out as yet another troublesome pest.

There has been much said about the content of oxalic acid in greens including sheep sorrel. The leaves are so tiny one would be hard pressed to eat enough to cause any issues. Oxalic acid is also prominent in spinach, peanuts, beets and chocolate. If you have kidney stones you may want to avoid grazing for an entire day on sheep sorrel leaves!

Take a look at some of the medical properties of this little powerhouse wild plant:

- Depurative-Supports kidney and liver function by cleansing waste and toxins, increases blood flow in tissues and encourages lymph drainage.
- Diaphoretic-Induces sweat aiding in toxin

removal.

- Antipyretic-Reduces fever and helps control body temperature.
- Anti-inflammatory-Reduces inflammation.
- Antibacterial
- Loaded with antioxidants

Sheep Sorrel also is very high in Vitamin C (1 cup of the greens equals about 80% of the daily recommended amount) and Vitamins A, K, and B-Complex. Also loaded with minerals including potassium, magnesium, phosphorus and zinc, it's no wonder that this is one of the four herbs used in a traditional, herbal anti-cancer formula (the other herbs are Burdock Root, Slippery Elm Bark, and Indian Rhubarb Root).

Whether you are foraging Sheep Sorrel for food or medicine, you won't be disappointed in this small plant with its unassuming look.

Wild Chamomile-Matricaria Chamomilla

Similar in smell and medicinal value to Roman Chamomile and German Chamomile, the wild version of this little flower can be found on fence lines, roadsides, open fields and very likely your driveway. It loves sunny, open areas and offers a natural way to find relief from stress related problems – simply by having a cup of Chamomile Tea.

There are two major differences between Wild Chamomile and the German and Roman varieties. One is that the flowers of the wild plant do not grow the white petals usually seen on the cultivated versions. Only the large, greenish yellow middle of the flower is seen. The second difference is that the wild version rarely grows higher than 6 or 7 inches at most. A ground-hugging herb, Wild Chamomile is easiest to identify by smell – the flower has a distinct Chamomile odor that is unlike any other flower's aroma.

Wild Chamomile can be eaten or made into a tea raw, or dried for later use. Harvest it by cutting the leaves and flowers. By hanging it upside down in an airy, sunless spot or lying out on a screen, it will dry rather quickly.

Internally, when taken raw, or in a tea or tincture, this plant has been long valued as a natural, gentle sedative, headache reliever and tonic for stomach ailments. Externally, Wild Chamomile's strong anti-inflammatory properties offer relief for all wounds where inflammation is present. This humble little flower, often blending in with the grass, also contains small amounts of calcium, magnesium, potassium, fluoride, folate and vitamin A.

Wild Rose-Rosa Rugosa

Like many plants across North America, the Wild Rose originated in Asia and has been around so long that most people consider it a native plant. Often viewed as an invasive shrub, which frankly it is, this herb has an incredible fragrance, beauty and so much medicinal value that it's certainly worth getting pricked by thorns to forage it's petals and hips.

The petals of the Wild Rose are a delight to eat raw and have been made into Rose Water for centuries. This invigorating herbal water, with its strong astringency and a light, sweet aroma, makes a refreshing, toning facial wash. When harvesting the petals, don't take every petal from a flower, leaving color for the bees to find them. Dry the petals by spreading them out single layer on large trays. Their size will be reduced significantly when dried, so pick plenty; usually where Wild Rose grows, there is abundance. As always, when working with prickly plants, wear sturdy long sleeves, pants and gloves if desired.

With more than 20 times the amount of Vitamin C as

oranges and when taken regularly, Wild Rose Hips (or the bulbous seed pods) work well in preventing colds and flus. Also present are antioxidant, astringent, anti-viral and diuretic properties. These medicinal actions have a positive effect in urinary tract infections and diarrhea.

The hips are traditionally preserved in syrups or jams. But fresh or dried rose hips can also be eaten raw or infused in tea. Either alone or in combination with other herbs, they make an aromatic and Vitamin C-packed drink. With a high content in iron, they work well in regulating menstruation.

Yellow Dock-Rumex Crispus

Another hard to miss wild herb, Yellow Dock can be found in any open areas, sprouting up in roadsides, gardens, fields and gravelly spots. All parts of the leaves, stalks and seeds are edible. In the spring and early summer, the leaves will be long and pointed and a green seed stalk will begin to grow. By late summer the seed stalk has turned a rich brown. The brown crispy seeds can be ground up and used as a flour substitute.

The leaves of Yellow Dock are high in oxalic acids. Eating it profusely is not recommended but is fine for occasional salads and teas. For someone with kidney disease, Yellow Dock should be avoided.

The real power of Yellow Dock, like many herbs, lies in the root. Bright yellow and extremely bitter smelling, the root provides a rich source of iron that is helpful for those with anemia. The root, which can be sliced, dried and ground, or tinctured, also works well with skin conditions, is a mild laxative, stimulates bile production and is a powerful antibiotic. Yellow Dock root when taken internally acts as a blood cleanser working its magic on skin disorders, chronic inflammation and respiratory issues.

When harvesting the root, you may need a large shovel to dig a circle around the plant before pulling up the long, thick root that can grow very deeply. The green seed stalks can be harvested by cutting off the top portion, steaming the greens or drying for later use in teas. These seed stalks will dry nicely while still in the ground and once they turn a deep brown color they are ready for preserving or for using ground up as a flour alternative.

CHAPTER 2

CULTIVATED HERBS GONE WILD

Medicinal Herbs You Can't Kill

Sometimes it's a fine line between a wild herb and a cultivated plant. At what point does a cultivated herb, after self-sowing itself for years, and with no care except for an occasional harvesting, become wild? There are plants that were originally cultivated that have been spreading themselves around now for centuries. Though they may have been introduced to an area by direct cultivation, after years of self-sowing and surviving they certainly look and act like a wild plant. In this Chapter, four medicinal herbs that once planted can easily take on their own survival, are explained in detail as to health benefits they provide and the ways in which they grow best.

Borage-Borago Officinales

This beautiful flowering plant, a favorite among bees, boasts gorgeous, almost neon purple flowers making it a popular ornamental plant and a colorful, exotic addition to salads. Once Borage is planted in your yard or garden, there will be no need to plant again as it will self-sow easily.

The leaves and flowers of Borage boast a high content of Vitamins, primarily C, and B3, and many minerals including iron, calcium, phosphorous and magnesium. The flowers are tasty eaten raw and can make a nutritional and beautifying addition to your dinner salad. Steaming the leaves and the flowers is another way to eat them

Medicinally, Borage has a wide range of uses. The leaves and flowers are used in a tea for its diuretic, sweat inducing, blood purifying, and sedative properties.

For these reasons, Borage is helpful for fevers, coughs, colds, bronchitis, skin disorders and depression. This plant as a whole contains high amounts of linolenic acid, which assists our bodies in producing GLA's, which are important omega 6 fatty acids. Due to the presence of linolenic acid, it is used to treat arthritis, providing 240mg of GLA's per 1000mg of the herb. The high content of mucilage in this plant often provides relief for digestive disorders.

Though Borage loves to self-sow in grassy areas, it particularly likes rich garden soil and once planted, it will appear throughout the garden, attracting bees and giving you an herb you can readily eat wild for its nutrients or take advantage of its healing properties.

Calendula-Calendula Officinales

Commonly known as Pot Marigold, this herb has flowers that can be eaten raw (one of the more delicious flower buds) or used to give a little sunny sparkle to your salad. When the flowers are left in the fall, their large seeds dry right on top of the plant, making it easy to save seeds by plucking them off. It's also possible to just let them fall and replant for next year. No additional care is needed to keep these going.

This modest little plant is widely known as one of the best herbs in treating skin conditions. Its vulnerary properties, which aid in new cell growth proliferation, along with its strong antiseptic, astringent and anti-inflammatory attributes, help external wounds and skin disorders heal quickly.

The tea of dried or fresh Calendula can be drunk to benefit from these actions or the cooled tea may be applied by soaking a clean cloth and placing directly on the affected skin. Cooled Calendula tea from a sterilized cup is effective in treating conjunctivitis and other eye inflammations. By eating the flowers raw, you can take advantage of the same actions as when drunk as a tea.

Making a sun infusion with oil and Calendula flowers is very easy and will be discussed in Chapter 5 along with information about using the infused oil to make healing salves.

In recent studies, the powers of Calendula have come to light in tests showing its effectiveness as a reliever of pain due to radiation effects in cancer treatment.

Peppermint-Mentha Piperita

Trying to eradicate this herb, once planted, may prove futile. It's best to just go for a bit of containment by giving Peppermint its own spot, away from other plants, and then make the best of this gentle and widely used herb. It doesn't take long for Peppermint, or any mint for that matter, to start acting like a wild plant, growing profusely, with its underground tendrils at times traveling under your driveway to the other side!

One of those tough plants that like to be mowed, Peppermint leaves and flowers can be eaten raw, made into a fresh tea, tinctured, or infused for hot tea. It makes a tasty addition to homemade lemonade. While Peppermint has only traces of vitamins and minerals, including Vitamin C and manganese, its acts as a gentle digestive stimulant, a nervine, and has carminative actions that make it perfect for stomach upset of any kind. This plant is effective in treating nausea, constipation, loss of appetite and even morning sickness.

Its aroma is said to increase mental power and memory and the herb is also used in treatment of respiratory ailments. Used with other herbs, Peppermint acts as a stimulant in the digestive tract and often causes an increase in the powers of the other plants.

With its abundant growth, Peppermint, even in

northern climates, can be harvested 3 or more times per season. You don't have to worry about cutting too much as this medicinal herb will give you plentiful leaves with which to work. It will grow even better if you make a cutting before flowering. Peppermint does not depend on its flowers for propagation because of its root stems that travel far and wide underground. The flowers do attract pollinating insects so it's good to let some of the plant flower.

Common Thyme-Thymus Vulgaris

This weedy little survivor, when planted in a rock garden will spread quickly and robustly, offering its many uses in cooking and healing. By simply planting a seedling in your garden (like Peppermint, you may want to have a site set aside), you will have many years of this herb to enjoy.

The volatile oils of Thyme, primarily Thymol and Carvacrol are found in all aerial parts of the herb and have very powerful antibacterial and antifungal properties. Since the oils are excreted through the lungs they provide strong actions when used for bronchitis and other respiratory diseases, including whooping cough. The volatile oils also act as an antispasmodic and an expectorant easing not only lung disorders but digestive system problems as well.

You can take advantage of all these benefits simply by eating some thyme, or making a tea with the fresh leaves and flowers, or by drying it for future use in teas, soups and almost any cooked dish. Like Peppermint, Thyme can be cut multiple times per season and grows profusely. Externally, baths with infused Thyme added in are used to soothe arthritic pain.

Because Thyme's volatile oils are so powerful, they may actually be toxic in large amounts. For this reason, Thyme should be saved for the conditions above and not used for chronic situations over a length of time. Pregnant women especially should seek advice from their healthcare practitioner before using Thyme medicinally.

CHAPTER 3

HOW TO HARVEST THE ROOTS, FLOWERS AND LEAVES OF MEDICINAL HERBS

There are various methods of harvesting the different parts of wild medicinal herbs. This chapter will outline for you the ways to dig, cut, and pluck the powerful plants in the wild, in a way that ethically preserves those plants. As always, be certain of what plant you are harvesting by having someone in your area that is familiar with local wild plants help you make a positive identification.

Roots

Generally, roots have the most medicinal power of all the plant parts. The two times when roots are usually harvested are in the spring or fall; this is when the aerial parts of the plant either haven't grown yet, or have died down, ensuring that the full strength of the root is present.

After you've made certain you are harvesting the desired herb, make sure there are plenty of other plants in the area; never take the only one. Start by loosening up the earth with a shovel around the plant, digging a wide circle around its base. Then harvest the root by gently pulling out the whole plant while holding onto its base,

Some herbal roots, like Yellow Dock, may take a little elbow grease to get it out. The easiest way to ensure success is to get the ground nice and loose all around the root. The ease of pulling it out also depends on whether the ground is wet or dry, sandy or loamy. In any case it will come out with a little work.

Let the root dry in the open air and sun for a few hours if possible. That way, once dry, it is easier to brush off the dirt. Using a clean cloth (that you don't mind getting dirty) gently rub the dirt off making sure to get in all the crevices. A vegetable brush would work well here. After brushing off the soil, wash it well. Then slice it as thinly as possible for optimal drying. If you have an oven that can be set at 200°F, spread the thinly sliced roots onto a cookie sheet or pan for an hour or two. Or, drying the roots in a dry, airy spot inside your home will work, just taking up to a week or so to dry completely, depending on the weather in your region. Harvesting roots in the fall improves drying conditions for those living in climates with wet, rainy springs.

Once the roots are dry and crispy they can be stored in a glass jar for using later to make herbal decoctions (discussed in Chapter 4, basically boiling the root for 20 minutes), or ground up into a powder and then stored. Powdered roots are often made into poultices that act as healing compresses, which are used externally.

Flowers

With the many different kinds and sizes of flowers, it's helpful to follow a few general guidelines for harvesting the different kinds of medicinal flowering herbs.

Some flowers, like those of Yellow Dock, Sheep Sorrel and Plantain, are too tiny to pick individually. Harvesting the entire seed stalk is much easier (as long as there are plenty of other seed stalks in the area) and makes it a quicker way to pick a quantity of small flowers. By cutting at the base of the flower stalk and drying it upside down you can preserve most of the flowers. Leaving a tray or plate underneath will catch any loose flowers that fall.

Other flowers are large and self contained enough to be easily plucked. Red Clover and Dandelion flowers easily detach and make for quick wildcrafting. They can be used in salads or teas either fresh or dried, and in wine if you can collect a lot of Dandelions. Both these flowers, and others that are dense, can be cut and left in the shade outdoors so the tiny bugs that might be inhabiting them can crawl away, which they usually do once the plant has been disturbed. If drying flowers, break them up to make for quicker drying time. Never pick an entire patch. It's important to leave flowers to attract bees and encourage self-sowing.

Flowers with large petals, like those of the Wild Rose, can be harvested by the petal, making sure to leave a few to attract the bees for pollination. These types of flowers often have more oils in them and can take a bit longer to dry. They also will often exhibit a marked reduction in size after being dried.

Some flowers, like those of the Wild Rose, Borage and Calendula, are tasty to just eat raw. Others are more commonly dried for medicinal and nutritional purposes, as well as for beautifying body care products.

Leaves

The harvesting of wild medicinal leaves is easy. Simply cut each leaf at its base or cut a bunch at once. The general rule is that leaves are best when young but still are usable even when old; they just are usually bitterer and their properties may not be as potent.

To dry leaves, lay them out single layer on a screen or tray. Thicker leaves like those of mullein or borage may be cut into smaller pieces to enhance drying time.

CHAPTER 4

HOW TO INCORPORATE WILD MEDICINAL HERBS INTO YOUR LIFE

Wild Greens for Supper

There is nothing like the feeling of being able to wildcraft greens from a source you know. You can eat a few as you gather them up, and after adding a few fresh picked flowers found on the way, you can return home to make your dinner more beautiful and nourishing.

Great examples of salad-friendly herbs include the leaves of Sheep's Sorrel, Dandelion, Peppermint (though usually just a few Peppermint leaves will add enough flavor) and the flowers of Red Clover, Dandelion, Wild Rose, Borage and Calendula. There are many more edible and medicinal herbs in the wild; this book introduces some of the most easily recognizable plants and once you have learned to forage and make use of *them*, you'll be ready to access the wealth of information on the web and hopefully learn more from other foragers as well.

Adding Sheep Sorrel leaves to your food gives a pleasant tangy, sweet and sour flavor. In salad they combine well with a simple citrus dressing or may be eaten plain. When eating an herbal and wild green salad, it's important not to douse it with a high fat, high salt dressing in order to enjoy all the medicinal and nutritional benefits.

Dandelion Greens, when eaten young, make a great bitter addition to the evening meal. In raw salads, the tender leaves pack a whopping mineral content. The older leaves, present after the plant has flowered, are much bitterer and may be better used in soups, smoothies or sautéed and added to stir-fries. The bitterest leaves can be steamed and tempered with a little lemon juice.

Remember, bitter greens are good for our bodies. Our digestive system will receive them thankfully as the greens not only provide important phytonutrients, minerals and fiber, but promote digestion acting as cleansers and tonifiers.

Peppermint not only dresses up a salad and provides an excellent source of fiber; this plant acts as an all around digestive tonic, perfect for mealtimes. By throwing a few Peppermint leaves into your pre-meal salad or by enjoying the few leaves garnishing your dessert, you can enjoy the improved intestinal efficiency that these beautiful leaves contribute.

A Calendula flower has many long thin petals and when plucked individually, can look as beautiful as saffron on your salads and main dishes. Not only do they make an appealing garnish (often times in a restaurant the healthiest item on a plate IS the garnish!), but also your body will receive the benefit of the high content of antioxidants contained in these little buds. Borage flowers may be used in the same way as Calendula and both kinds of flower petals are a beautiful way to top a soup, stir-fry or salad.

Some people like to deep-fry their flowers and this may be delicious but it sort of defeats the purpose of eating wild medicinal plants in the first place!

All of the above mentioned herbal leaves and flowers could simply be picked and eaten raw, added to other greens, eaten alone in a salad or thrown into any cooked recipe. Give your meals a touch of color and a bunch of mineral-rich greens from the wild and reap the many healthful benefits.

Drying Herbs for Later Use

Eating wild medicinal herbs and greens raw is a wonderful way to commune with nature, fortify your body and enjoy sunshine and fresh air. But for those of us living in colder climates, foraging gets a bit tricky when the ground is frozen solid and covered with three feet of snow. For this reason it is wise to learn how to dry and preserve wild herbs, because a long winter can test our health and having wild medicinal herbs saved in the pantry will offer many rewards.

When drying the herbs you have harvested, it's good to lay them outside in the shade for at least a few hours setting them on a tray or plate or in a basket. This lets excess moisture evaporate and makes drying easier and quicker and reduces the chances of your herbs mildewing before completely drying. Once brought inside the herbs can stay on their trays in an area that is warm and well circulated and out of the sun.

Some herbs have a higher content of oils or mucilage and take longer to dry. (i.e. mints and Mullein) Generally the herbs with high tannin content (like Raspberry, Thyme and St. John's Wort) are the quickest to dry. You'll know your plant is dry and ready for storage when it crumbles crisply. Once completely dry, the herbs can be transferred to a paper or canvas bag or a glass jar (the next chapter discusses at length storage methods). Do not let your herbs stay on their trays longer than needed; they will collect dust and their medicinal power will wane faster than when properly stored.

If you live in a humid climate, you may want to keep the drying herbs in a warm spot with a fan in the room. If an herb is not drying and is keeping its water content, it will soon start to turn brown or black on the edges. At this point you should compost these herbs and not save them. If your climate is too moist, or as it often happens in some coastal areas, the time to harvest herbs is also the dampest time of the year, using a dehydrator or an oven set on 200°F for a few hours will do the trick.

Storing Harvested Medicinal Herbs

Once you have dried the wild plants you plan to use later, a sterile glass jar with its lid is probably the best and most environmentally sound way of storage. A colored glass jar is even better, as it will preserve the color of flowers for longer.

To sterilize your glass jars simply dip them into boiling water and take them out (you can use tongs, or a spoon and a butter knife works well) and let them dry completely mouth side up. It won't take too long to dry because they will be so hot from the boiled water.

A cool, dark pantry is probably the best place to store your dried herbs, but a kitchen cupboard will do nicely as well. If you have had a particular medicinal plant stored for a long time and want to know if it's still viable, asking yourself a few questions and using your senses will help determine whether or not the herb is still good. Does the herb still have its strong smell? Is it the same color as when you originally stored it? Over time the smell and color can be leached out giving you a good signal that it is time to compost these herbs. They probably can still be used but they won't have their full power. If you ever notice that a dried herb feels at all damp or smells even just a bit mildewy, it is a sign that moisture has found its way in and it's time to discard that herb.

Making Herbal Teas

In cultures all over the world, offering a cup of tea to a guest is an ingrained part of social etiquette. Not only does it offer a pleasant way in which to chat with a neighbor or catch up with an old friend, but the warmth of the cup in itself and the healthy ingredients within can relax and fortify.

Making your own herbal tea, with wild plants that you've harvested and dried, is an art in itself. Are you in the mood for something aromatic and tonifying, like Rosehip Leaf Tea? Or how about a good dose of potent greens for a boost of mineral content? Maybe you're looking for a great after-supper tea, something relaxing and soothing to your digestive system. Then you may want a cup of Peppermint and Chamomile tea. An excellent spring tonic tea would contain leaves of Dandelion, and Sheep Sorrel . By matching moods and physical needs with the different herbs in your pantry, you create an individualized, medicinal cup of tea for yourself or your child, friend or loved one.

To make your cup of tea, when using any of the aerial parts of the plant, simply use a tea ball to contain the herb, placing it in a nice large mug. Pour boiling water over it, cover and leave for about 20 minutes, then enjoy! If you don't have a tea ball, you can place the herbs into the boiling pot of water, after taking the pot off the heat. Then cover and leave for 20 minutes and strain afterwards. It certainly won't hurt to eat some of the herb (you will find with Wild Chamomile that the smallest parts are usually floating in your cup).

If you are using a root, like Dandelion or Yellow Dock Root, you will need to *decoct* the herb to maximize its properties. This only means simmering them for 20 minutes, then straining and drinking your tea.

For a general measurement, you can add one small handful of herbs to the boiling water per cup of tea. After time your preferences and tastes will dictate how much you use.

How to Make a Tincture

There are three main solvents that are used when making tinctures, Alcohol, Vinegar and Glycerin. All three have their pros and cons and all three will make a satisfying extract with your wild medicinal herbs.

Alcohol is by far the most efficient way to extract the medicinal properties from a plant. Usually a 50/50 ratio is used, alcohol to water. By using 100 proof alcohol (which is 50% water) you may just use the herbs and the alcohol without adding any water. Just fill a sterile jar with the chopped, fresh or dried herb (fresh being the ideal, but dried herbs are certainly doable) and pour the alcohol over the herbs until full, making sure it completely covers all the herbal parts. Stems or petals or other parts that stick up out of the liquid may start to break down and contaminate the tincture. Seal with a new lid.

Now place the jar in a dark, warm spot for at least 2-4 weeks. A longer amount of time, even up to 6 weeks or more is desirable to some herbalists. The longer it macerates, the stronger the tincture will be, leaving you with no real time constrictions. It's good to shake your jar every few days so the herbs to do not get stuck at the bottom.

For straining the tincture, a large stainless steel strainer lined with 4-5 layers of cheesecloth may be used. With your hands squeeze the liquid out of the herbs, and then press out the remaining liquid by gathering up the cheesecloth into a ball and squeezing that.

For maximizing the potency and lasting power of your tincture, store in dark colored glass, or clear glass in a dark place. It helps to use new or recycled tincture bottles that have the dropper for easy dosing. But larger glass containers will work and can be dosed out using a 1/8 measured teaspoon. When using the dropper, about 15 drops is the standard dose.

Some tinctures made with tonifying herbs, such as Dandelion and Saint John's Wort, can be taken once every day for an extended period of time. Other herbal tinctures, like Yellow Dock, may be taken 3 times per day over a short period of time for acute situations.

When using dried herbs as opposed to fresh, you may want to add a bit more water to the extract since the fresh herbs have more water content to begin with.

Organic alcohol is available and makes a good choice for tincture making.

The tinctures made with alcohol are not always the best tasting, but they do have the reputation of being the most potent extracts. If the tincture is too strong to take by mouth, it can be mixed with a bit of boiling water and the alcohol will evaporate leaving the herbal extract in the water for drinking. If you really hate the taste of alcohol, or do not want to use any product with alcohol (even the 15 drops usually dosed may be unacceptable to an alcoholic) the following two solvents also work well and have many benefits. Alcohol tinctures can last years when stored properly. Always be sure to label your tincture.

Vinegar makes a strong herbal tincture, even though it does not extract quite as many plant constituents as alcohol. Many people prefer the taste of vinegar tinctures above the alcoholic preparations. As an added plus, vinegar acts as a natural pH balancer in our bodies and tonifies the digestive tract. It makes an excellent alternative for those who cannot drink alcohol. When making a tincture to tone a body system, vinegar works well for this purpose.

The instructions for making the vinegar tincture are the same for alcohol except warming the vinegar slightly first will help in the extraction process.

Glycerin

Glycerin is the sweet, mucilaginous part of fats from animals and plants and is available in natural food markets and online. Its sweetness makes it an appropriate medicine for children.

Glycerin by itself is nutritive and will act also as a preservative. Again, alcohol will extract the most out of your wild medicinals but glycerin will make a satisfactory tincture.

When making a glycerin tincture, use 4 parts water to 1 part glycerin and be sure to use 100% vegetable glycerin. Then follow the instructions for making and storing an alcohol preparation.

Tinctures are arguably the best way to get the most

out of your wild medicinal plants and learning to make and store them in this way ensures you have these powerful plants available to you all year long.

Herb Infused Oils

When an herb is infused in oil, medicinal properties are extracted and then the oil may then be used in salves, massage oils and other herbal preparations. Herbal infused oils are not to be confused with 100% Pure Essential Oils which require distillation. There are two types of infused herbal oils discussed here: sun infused oils and stovetop infused oils.

To make a sun infused oil use either dried herbs or fresh wilted (using fresh, with their high water content will cause bacterial growth in your infused oil and eventually ruin it).

To fresh wilt your foraged plants, pick them in the morning and lay them outside somewhere warm and shady, leaving them there until dark, before dew takes place. Then fill a glass jar to two thirds full with your dry or fresh wilted herbs and completely cover with the oil of your choice.

Olive Oil, Almond Oil, Grapeseed Oil and Apricot Oil all make great options. Make sure no herb pieces are sticking out of the oil; they must be completely covered by the oil to ensure no decaying happens. Cover tightly and put in the warmest spot in your yard or deck and keep it there for between 2 and 4 weeks, shaking it up once in a while. Some folks will submerge a quarter of the jar in sand or soil to help keep it warmer. Then strain with layers of cheese cloth and store your sun infused oil in a glass jar. A cool, dark environment will work best; the colder the storage area the longer they will last.

For the Stovetop Method of making infused oils, again you may use either dry or fresh wilted herbs. A double boiler will work best because you do not want your oil to boil. If your stovetop has a low enough setting to heat the oil without ever simmering it, then that's an appropriate method as well. The double boiler will make this easier. Boil the water and place the herbs in the top part, making sure to completely cover with oil and **heat** while covered (not boil or simmer) for 45 minutes.

After 45 minutes the herbs can be strained and stored using the same method as the sun infused oils.

Herbal infused oils can be added to your bath, made into creams and salves or used for massage.

CHAPTER 5

TEN MEDICINAL RECIPES FROM THE WILD

In this guide you have been introduced to wild medicinal plants and how to forage them and what they are used for. Now it's time to create your own herbal medicine cabinet with the wild herbs you foraged. The following selection of recipes was developed to introduce you to as many methods of herbal preparations as possible using the very herbs you learned about in the previous chapters.

Solar Infused Mullein Oil

Ingredients:

- 1.5 cups of Olive Oil (or oil of your choice)
- 1 cup fresh wilted or dried mullein flowers

Directions:

Prepare by filling a jar 2/3 full with the herb and covering with oil. Use more oil if needed to ensure all the herb is covered completely. Let sit in sun for 2-4 weeks, strain with layers of cheesecloth and store in a glass jar in a cool, dark place.

To use:

Warm the oil slightly, let it cool to room temperature and place a drop or two in affected ear 3 times per day.

Calendula Healing Salve

Ingredients:

- 5 parts Solar or Stovetop Infused Oil of Calendula
- 1 part Beeswax (Candellia Wax is a vegan option, but use a slightly smaller amount than beeswax)

- 1 teaspoon Vitamin E Oil (preservative)

Optional: 100% pure essential oil of your choice. Lavender and Tea Tree are good options. 5-20 drops depending on amount of salve you are making.

Directions:

Melt beeswax and oil together in a pan, add optional essential oil and pour into container of your choice. Let solidify and cool, then put lid on and it's ready to go.

To Use:

Apply directly to cuts, scrapes, hangnails, splinters (helps to draw), eczema, diaper rash, wounds, chapped lips, minor burns or any other skin condition.

Saint John's Wort Capsules

Ingredients:

- Dried and ground Saint John's Wort herb

- Vegetable capsules (which can be easily ordered online or bought in natural food markets)

Directions:

Simply fill your capsules with the dried herb and close. Store the capsules in a recycled and washed vitamin jar or any other container. Refrigeration will make them last longer. Also available are little capsule trays that allow you to make 50 or 100 capsules at once rather quickly.

To use:

Take 1-3 capsules daily, especially good for going through life changes and stressful events.

Dandelion Root Coffee

Ingredients:

- Roasted Dandelion Root (to roast, wash fresh roots and then chop them into many small pieces; roast in shallow pan or cookie sheet at 250° for 2-3 hours or until they are a nice chocolate color. Stir root bits every 20 minutes or so)

- Water

Directions:

For each cup of water, add in 1 Tablespoon of roasted Dandelion Root. Boil the herb and water for 20 minutes, strain and enjoy.

To use:

Makes an excellent spring warm-up and detoxifier. Add sweetener of your choice if desired.

Red Clover Tea

Ingredients:

- Red Clover, fresh or dried
- Water

Directions:

Bring a pot of water to boil. Add 2-3 handfuls of fresh or dried Red Clover blossoms and remove from heat. Cover and let infuse for 20 minutes. Strain and drink.

To Use:

An excellent and soothing remedy for colds and flu.

Yellow Dock Tincture

Ingredients:

- Fresh or Dried Yellow Dock Root pieces chopped
- Organic 100 Proof Vodka

Directions:

Place dried root pieces into a sterile jar filling it 2/3 full. Add alcohol all the way to the top completely covering the herbs. Cover with a new lid and let macerate in a dark, warm place for 4 weeks, shaking it every few days. Then strain, rebottle and store in cool, dark place, preferably in colored glass.

To use:

A great blood cleanser, take 15-30 drops per day to treat chronic skin conditions.

Totally Wild Herbal Salad

Ingredients:

- Sheep Sorrel Leaves
- Plantain Leaves
- Dandelion leaves and flowers
- Optional: Juice of one Tangerine

Directions:

toss all the leaves together, top with broken up Dandelion flowers and add any optional ingredients (good choices may be other salad greens, raw vegetables, raw nut slivers and sunflower seeds) and drizzle free squeezed Tangerine juice over the top.

To Use: Just eat it!

Instant Boo-Boo Reliever

Ingredients:

Wild Plantain Leaves

Directions:

Pick the leaves and chop them or if in the wild, simply mash them up as best you can with your hands.

To Use:

Apply directly to any kind of external wound, keeping on as long as necessary. Cover with a bandage and leave on for a few hours. Repeat as necessary.

Chamomile Bath Salts

Ingredients:

1 cup Epsom Salts

- ½ cup (or less) Sea Salt
- ½ cup ground, dried Wild Chamomile
- 20-30 drops of essential oil of your choice (Rosemary is nice, or Lavender)

Directions:

Mix all ingredients together, stir well and store in glass container. Use ½ to 1 cup per bath. If you don't have a tub, this recipe works wonders in a foot bath.

To Use:

Simply add to your bath to enjoy a spa-like relaxing treatment

Wild Rose Water

Ingredients:

- 2 Cups Fresh Wild Rose Petals

- Water

Directions:

Place Rose Petals in a sauce pan and add enough water just to cover. Simmer about 30 minutes or until the petals have paled. Then strain and store in a glass jar in a cool place.

To Use:

Can be put into a spray bottle as a spritzer to freshen the air or your skin, added to your bath, or used to splash directly onto your face as a morning wash.

CONCLUSION

EMBRACING THE POWER OF PLANTS

Even as the health of our global environment is in a state of deterioration, it's inspiring to see wild green herbs growing up through cracks in the pavement. There's something hopeful about a Saint John's Wort plant sprouting along the edge of a busy parking lot or Wild Chamomile spreading freely all over your driveway – and in this stressful world, it seems as though they are just patiently waiting to be picked. Waiting to join us on our journey to healthier living and wellness.

While gathering and using wild medicinal plants you can feel a greater connection to this earth while getting a chance to experience the gentle healing actions of the roots, leaves and flowers. Touch the ground, eat from the wild and feel better. There's no mistaking it – there is power in medicinal herbs.

Remember to always practice foraging and herbalism with respect. The powers of wild plants can be very strong and it is essential to consult with your health care practitioner before enjoying the power of plants. Then ask a local to introduce you to the herbs in person.

Lightning Source UK Ltd.
Milton Keynes UK
UKHW020650041220
374589UK00007B/139